I0797698

RATTLESNAKES

by Matt Lilley

Cody Koala
An Imprint of Pop!
popbooksonline.com

abdobooks.com
Published by Pop!, a division of ABDO, PO Box 398166, Minneapolis, Minnesota 55439.

Printed in the United States of America, North Mankato, Minnesota.

052021
092021

Cover Photo: Shutterstock Images
Interior Photos: Shutterstock Images, 1, 6, 19 (bottom left), 19 (bottom right); iStockphoto, 5 (top), 5 (bottom left), 5 (bottom right), 10, 12–13, 19 (top), 20; Rick & Nora Bowers/Alamy, 9; Bill Gorum/Alamy, 15; Photography by Adri/Alamy, 16–17

Editor: Aubrey Zalewski
Series Designers: Laura Graphenteen and Colleen McLaren

Library of Congress Control Number: 2020948904
Publisher's Cataloging-in-Publication Data
Names: Lilley, Matt, author.
Title: Rattlesnakes / by Matt Lilley
Description: Minneapolis, Minnesota : Pop!, 2022 | Series: Desert animals | Includes online resources and index.
Identifiers: ISBN 9781532169731 (lib. bdg.) | ISBN 9781098240660 (ebook)
Subjects: LCSH: Rattlesnakes--Juvenile literature. | Snakes--Juvenile literature. | Reptiles--Juvenile literature. | Desert animals--Juvenile literature.
Classification: DDC 591.754--dc23

Hello! My name is

Cody Koala

Pop open this book and you'll find QR codes like this one, loaded with information, so you can learn even more!

Scan this code* and others like it while you read, or visit the website below to make this book pop.

popbooksonline.com/rattlesnakes

*Scanning QR codes requires a web-enabled smart device with a QR code reader app and a camera.

Table of Contents

Chapter 1

Desert Rattlers

Many kinds of rattlesnakes live in the deserts of North and South America. These snakes have rattles on their tails. They also have sharp fangs with **venom**.

Watch a video here!

rattle

The snake's rattle is made of loose scales. When the snake shakes its tail, the loose scales knock together. This makes a rattling sound.

Chapter 2

Hunting

Rattlesnakes eat animals such as mice, rabbits, and lizards. The snakes wait for their **prey** to come close. Some rattlesnakes wait for weeks to catch any prey.

Learn more here!

pit

Rattlesnakes can sense prey in the dark. They have pits on their faces. Snake pits look like holes. With their pits, they can see animals' heat. This helps them find prey.

Snakes do not have eyelids. They have clear scales over their eyes.

body
pit
rattle

Rattlesnakes use fangs to bite their prey. They strike quickly and let go. The **venom** kills the prey. Then the snakes eat the prey.

Chapter 3

Life in the Desert

The desert gets hot and cold. On hot days, rattlesnakes rest in **dens**. On the hottest days, they only come out at night. In winter, rattlesnakes **hibernate**.

Learn more here!

The desert is dry.
Rattlesnakes get some water from the animals they eat. They also get it from the rain.

When it rains, they curl up.
Their scales collect the water.
Then the snakes drink it.

Chapter 4

Life Cycle

Rattlesnakes **mate** in the spring. Rattlesnakes give birth to approximately ten live babies. The mother does not take care of her babies.

newborn rattlesnake

Complete an activity here!

rattlesnake skin
after molting

Baby snakes **molt** after a few weeks. They shed their skins. They grow bigger. After their first molt, they leave the area they were born in. Rattlesnakes can live up to 25 years.

Baby rattlesnakes do not have rattles at first. They get rattles after they molt.

Making Connections

Text-to-Self

Rattlesnakes can see body heat. What do you think that would be like?

Text-to-Text

Have you read about other animals that live in the desert? How are those animals similar to rattlesnakes? How are they different?

Text-to-World

Rattlesnakes have fangs with venom. Can you think of another animal that has venom?

Glossary

den – a cave or hole that shelters an animal.

hibernate – to enter a resting state during winter.

mate – to come together to have babies.

molt – to shed an outer layer.

prey – an animal that is hunted by other animals.

venom – poison from an animal bite or sting.

Index

Online Resources

popbooksonline.com

Thanks for reading this Cody Koala book!

Scan this code* and others like it in this book, or visit the website below to make this book pop!

popbooksonline.com/rattlesnakes

*Scanning QR codes requires a web-enabled smart device with a QR code reader app and a camera.